Work for the Tree

Jue Chang

美商EHGBooks微出版公司
www.EHGBooks.com

EHG Books 公司出版
Amazon.com 總經銷
2019 年版權美國登記
未經授權不許翻印全文或部分
及翻譯為其他語言或文字
2019 年 EHGBooks 第一版

ISBN-13：978-1-62503-535-6

Contents

Waking Up

Hello! Get up.
Hello! It's time to get up.
Someone shouted my name,
Someone is calling me,
Someone told me to get up.

I heard my name,
Instinctively responded,
A familiar name,
Although I have not woken up yet,
But I know that is my name.

Respond

My name,
My exclusive name,
Used to distinguish me and others,
Used to call me,
Yes, I am of this name.

I was called,
I know that the other person is calling me.
"What?"
"What's the matter?"
I responded to the calling.

Clock

"What's the matter?"
The other party did not answer me.
The other party just pointed to the clock.
Looking at the pointer,
I thought about it for a moment.

I woke up,
I thought of it,
There is something to do today,
I have something to do,
Get ready!

Money

What are you going to do today?
Same with yesterday,
Just eat well,
Hustling for food,
Work hard.

Work, make money,
Eat and buy bread.
The rice is from the land,
Bread is made by others,
Money is issued by the state.

Doorway

I am heading out,
Outside the door is a big tree.
A big, tall tree,
As high as the sky,
Look up and I will see it.

The roots of the big tree are very deep and thick.
The big tree grows food,
The big tree grows bread,
The big tree grows money,
The big tree is the country and the world.

Lunch

My job is to take care of this big tree.
Everyone has to take care of this big tree.
Transport nutrients,
Transport moisture,
Work like a cell.

Life makes sense,
Survival is accompanied by obligations,
I have something to do,
I want to eat,
I want to have lunch under the tree.

Opening an Account

Having lunch,
The tree called my name,
Lunch following the name,
The tree doesn't know me,
Only knows my name.

Although I am not necessarily a celebrity,
I still have to cherish my name,
The big tree is like this,
Name comes first,
People second.

The First Job

Taking care of the big tree is my job,
If the big tree is wrong,
If the big tree hurts itself,
I want to tell it.

Big tree!
The name is not a cold code,
The name is a warm person,
Please do remember.

If the big tree forgot, I will remind it again.
This is my job.

Function

Working every day,
Serving people,
Providing labor for the society.
Because there are many people at work,
The functioning of society is working.

The sun rises every day,
I want to live,
The big tree is also living,
The cell is working,
Rotation without interruption.

Family First

Work is to make money,
Work is to support the family,
Family is the most important thing,
Because the family needs to eat,
So we work.

Family is the most important thing,
Family members have priority over the state,
When politicians encourage sacrifice for the country,
Think about it,
Is it really that dangerous?

Working Hard

Work is two-way,
Give labor,
Harvest money,
Money went into the family,
Where did the labor enter?

Labor entered the market,
For the needs of others,
I have served others,
Others serve more people,
We all work for the big tree.

Still

What do people do by living?
What do people do coming to this world?
What is the world doing?
The sun did nothing,
The earth is always in a circle.

What is the big tree doing?
The big tree did nothing,
The big tree is breathing,
The big tree is growing up,
The big tree stays still.

Breathe

The big tree breathes quietly,
Thousands of years, tens of thousands of years,
In the long time,
In fact, there is nothing special to pursue,
The big tree simply survives.

The big tree breathes quietly,
In a long time,
The function is maintained,
People are running around,
Survival is a busy thing.

Falling Leaves

There are many things to do in life itself.
Food, clothing, living, traveling...
The hustling is already reduced by working together,
I have already pursued convenience through exchange.
But still very busy.

The work is like this,
Metabolism,
There are new leaves every day,
There are always leaves to sweep,
Work can never be done.

Clean

Life is the subject,
Life is the answer,
What I want is to live well,
What the people want is to live well,
The big tree is the same as me.

The big tree is very quiet.
I want peace,
The big tree is very clean,
People keep it clean and tidy,
I keep it clean and tidy.

The Second Job

I live under the big tree,
The big tree needs someone to clean it.
I need a place to live,
Let me clean the tree!

There are often fallen leaves,
Occasionally pests,
Is the tree healthy?
I have a lot of work.

Maintaining a place to live,
This is my second job.

Sunny Day

Sun rise,
I greeted a new day,
The big tree greeted a new day.
I have something to do,
The big tree has something to do.

Today is sunny,
I am stretching,
The big tree is stretching,
The hand reached into the sky,
The big tree stretched the branches and leaves to the sun.

Transportation

The more the tree grows, the higher
The distance between the roots and the leaves has become
farther.
Increasingly advanced technology,
Traffic getting more developed,
I can reach out to things farther.

I have transportation,
With the help of the machine,
I can be faster,
I can go further,
Science is great.

Competition

The big tree is very tall,
I am small,
After the science is developed,
The big tree is getting higher,
Am I getting smaller?

Is this a competition?
Is this hostile?
The big tree in front of me!
Do you want to be an enemy of me?
Do you want to be an enemy of yourself?

Truth

Technology cannot violate the truth.
The state cannot be an enemy of nature.
The state cannot be an enemy of the people,
This is a simple truth,
But the big tree is often forgotten.

Ok!
This is my job,
The state has the responsibility to care for the people,
The people have a responsibility to remind the country,
Don't forget the truth!

Shine

The big tree is big,
I am small,
What is bigger than the big tree?
Bigger than the big tree is the sun,
Bigger than the big tree is the truth.

I can't call the big tree to listen to me.
I can't ask someone to listen to me.
Sun!
Please lend me strength,
I want to make shine bright.

Reason

What do people do coming to the world?
Life makes sense,
I have work to do,
I want to learn the truth,
I want to learn to be a person.

I have to learn,
I want to reason,
I want to reason with the big tree.
I want to reason with others,
I want to make more people reason.

Throw a fist

Technology makes people lost,
Money makes people lost,
Power makes people lost,
Big tree and I both are,
This is the process of growing up.

When I was a child, I always cried unreasonably.
And threw a fist with the people around me,
Threw a fist with myself,
Big tree!
Let's get to know the truth together!

The Third Job

What do people do coming to this world?
Of course, learning to be a human,
Of course, learning the truth,
Learn with the big tree.

The big tree is still learning,
I am still learning,
Maybe I should back up temporarily,
But the truth is still the truth after all.

Learning to be a human,
This is the work to be done every day.

Source

If the faucet has no water,
If the bank goes bankrupt,
If the law is not followed,
If everyone is crazy,
How can I live?

There is a source in wealth,
It is the " Lender of Last Resort ",
It is the country,
Because of the shelter of the big tree,
People have a stable life.

Reliance

I rely on big trees,
The big tree also relies on me.
There is another source of wealth,
That's me,
That is the sun in my heart.

If everyone does not want happiness,
Any system is useless,
I hope to have happiness,
People want happiness,
The sun in the heart is the ultimate reliance.

Sound

The earth accompanies the sun,
The big tree accompanies the people,
The big tree is breathing,
Absorbed the energy of the sun,
Some things are getting different.

The big tree is rooted,
The fruits grow out of the tree,
The people obey the law,
The law is sound,
Wealth began to accumulate.

Police

What am I going to do today?
When you open the door, you will see the big tree.
I want to work,
I want to make money,
I have to pay taxes.

The tax became a policeman.
The law has an avatar,
The police patrolled,
The police are working,
The duty of the police is to protect the people.

One

The police are the law,
The police are the country,
The police are the people,
The police are individuals,
The police are a group.

The big tree breathes normally,
I need to eat,
The police have to eat too,
We have a common desire:
"I hope to live well."

Belief in survival

Is everyone afraid of death?
No, the belief in survival can be strong and weak.
Is everyone selfish?
No, some people don't care for life,
The belief in survival is a precious treasure.

Support each other!
Restrict each other!
The police laid the network,
"Please cherish life!"
The police said so.

The Constitution Law

Today is different from yesterday,
There is warmth in my heart,
Habits accumulate on the body,
"I have to live well."
I have to tell every cell I have.

Today is different from yesterday,
Faith is written into law,
The big tree must also obey the law.
"I have to live well."
Every cell tells the big tree like this.

The Fourth Job

What am I going to do today?
Survival is accompanied by obligations,
I want to obey the law,
I have to pay taxes.

The police need to eat,
The law needs to be maintained,
Who has the job of maintaining the law?
Yes, it is me.

Obey the law, maintain the law,
This is my job.

Dream Comes True

Survival is accompanied by the law,
Truth and law must be obeyed,
The big tree breathes normally,
Peace is the foundation of wealth,
After that peace?

If the dream comes true,
If the people are enthusiastic,
If the country is normal,
If there is only a small quarrel without a big war,
What should be pursued after peace?

Making money

What should I pursue after peace?
Of course it is making money!
Of course, it is economical!
People are going to eat when they are alive.
This is the eternal pursuit.

Peace is possible,
It is impossible to have no competition,
Although the big tree is normal,
The competition continues,
Ok! Nice to meet you.

Cooking Competition

Reaching the high peak,
Ultimate breakthrough,
Let's encourage each other!
Let's make progress together!
I hope this is a good game.

You bring out your dish,
I bring out my dish,
Taste each other's!
Find out the result!
How is it? Tasty?

Conquer

I work for the big tree,
I work for someone else,
Come up with better service,
Come up with better products,
This is work.

The product appears in life,
Fill the stomach of the guest,
The song fills the dream of the audience,
How is it? Is it hard to forget?
I used the work to conquer the big tree.

In instalments

The big tree is metabolizing,
Metabolism is not a matter taking for granted,
Must always fill up,
Must always come up with something new,
The cooking competition continues.

Take out new results every day,
A new day will bring out new results.
The days are the same,
Rack my brain every day,
Ah! This is work!

Answer Twice

The sun shines,
The big tree is metabolizing,
What is the blank today to fill in?
Big tree doesn't know how to answer,
Let me answer the big tree!

What are you going to do today?
This question has to be answered twice.
Big tree,
Thanks in advance to the sun!
Then start today's work.

Studying

It's a new day,
The cooking competition is still going on.
Let's eat together!
Let's study together!
Eating is endless learning.

What are you going to do today?
Nature has its truth,
But I shouldn't just obey,
I should take the initiative to pursue,
I want to show my results.

The Fifth Job

What are you going to do today?
Blanks are accompanied by hunger,
I will be hungry,
The big tree will also be hungry.

I will be hungry,
I want to make money,
The big tree will be hungry,
I want to produce results.

A new day will come up with new works.
This is my job.

Ancient Men

It's a new day,
The game with the big tree is still going on.
The big tree is really amazing.
The tree has lived for a long time,
The big tree has been through a lot of games.

It is strong,
Old tree,
It is strong,
Ancient great men,
However, I am very good as well.

Prejudice

It's another new day,
The game with the big tree is still going on.
The big tree still remembers the previous game.
What did the big tree bring out today?
The tree took out the wisdom of the ancients.

It's another new day,
The game with the big tree is still going on.
I know that today is a new day.
I can't be the same as yesterday.
I took out a new work.

Comparison

Life is always full of comparisons,
I compare with others,
I compare with myself,
I compare with yesterday,
I compare with the market.

Others are also studying,
Others are making progress,
There are always new products on the market.
The big tree is improving,
I am also making progress.

Beginner

The big tree is improving,
But big trees usually only slowly progress,
The big tree is a teacher,
The big tree is the returning champion.
Big tree has a successful experience.

The big tree is improving,
I have to improve,
I am a challenger,
I'm a beginner,
Beginners are the bravest ones.

Test

The teacher gave me the test,
I asked the teacher questions,
The exam makes it difficult for students to handle.
The student's questions make it difficult for the teacher to
handle.
Today is another new lesson.

The boss threw the problems to the employee.
The employee threw the problems to the boss.
How can there be so many problems,
The problem can never be solved.
Customers always ask new questions.

Troubles

The big tree has ancient wisdom,
Life is full of wisdom,
No, life is full of problems,
Solving them is called wisdom,
The unsolved ones are called troubles.

The big tree lives,
The big tree has encountered troubles.
The big tree is like a boss.
The boss also has a lot of things that he doesn't understand.
How to do it? Who will solve it?

Asking

There are many problems in life,
Ah! This is work!
I am here to solve problems,
The troubles that the boss can't solve,
Give them to me!

Boss!
Please remember to be polite,
Please listen,
If you are not quiet,
How can I help you solve the problems?

The Sixth Job

What are you going to do today?
There are always many problems in life.
What are you going to do today?
The big tree has a heart lock.

Big tree!
Today is the question,
Today is also the answer,
The answer is on you.

I am a student of the big tree, I am the teacher of the big tree,
This is my job.

Daily Routine

Live with the big tree,
There is wisdom in life,
This is a precious treasure,
Life continues,
What else is it to pursue?

What is the pursuit?
Life is the answer.
I am pursuing normal, ordinary.
I want to eat a meal with laughter.
How to achieve this wish?

Sportsmanship

Even if it is peaceful,
We still have to eat,
The game with the big tree is still going on.
Big tree!
This is the game not a war.

The table is ceremonial,
There are rules in the game,
The game has something other than victory and defeat.
I hope that you are a respectable opponent.
The spirit of sportsmanship is very important.

Play the Ball

I want to eat,
I want to laugh too,
People want peace,
People also want to sweat,
Let's fight without bloodshed!

Duel!
Ok! What is it?
Fighting is illegal.
Let's play!
See you at the stadium in the evening.

Sports Day

The game is not just a win,
There is sweat and laughter,
There are also crying, cheering, remorse, admiration...
The lion in the zoo is not happy.
What people want is not just to eat.

I want to eat a meal with laughter.
The big tree needs laughter,
The big tree needs vitality,
Let's play together!
Let's have a sports day!

Hero

If it is peaceful,
Is there no brave man without a demon?
Is this peace?
Without brave men and heroes,
Only the indifferent people.

The country still needs brave people.
Even if you are peaceful, you still have to exercise your mind
and body.
Heroes continue to pursue to be stronger,
Doing monotonous training day after day,
The hero went from the battlefield to the court

Bullying

Only the light can knock down the darkness,
Only the brave can defeat the devil,
Why is there a bullying?
Because of boredom,
Because of not knowing what to do.

The tree was once dead,
The tree once cursed the eternal long time,
The big tree doesn't know what to do alive,
The brave man challenged the big tree.
The brave man brought sunshine to the big tree.

Good Opponent

I have to face my life,
The big tree has to face the long time,
I look at the big tree,
The big tree looked at me,
We are good opponents.

Replace the war with the game,
This is true peace.
There is spirit outside the rules,
There is friendship outside of winning and losing.
This is true peace.

The Seventh Job

There is still the next game after another,
Repeating monotonous training,
Happy?
Ah! This is work.

This is work,
Happiness also needs training,
If I am not happy, I must train.
I want to make myself enjoy it more.

Repeat the same training every day,
This is my job.

Evolution

Repeat the same training every day,
Repeat and repeat,
Repeat and repeat,
I have made some changes,
The tree has changed a bit.

This is evolution!
Run faster,
Jump further,
Evolution is not just external,
Later, even my heart was changed.

Gene

I want to manage my heart,
I can't be willful,
I want to be a civilized person,
You can't be willful, big tree.
You have to become civilized, too.

The sun painted the trajectory,
What is civilization?
Civilization is a blueprint,
Civilization is the genes,
The big tree and I have become us.

Cultivate Deeply

What to do after peace?
Continue to compete with the big tree?
Have you been competing peacefully?
How stupid!
There are other things besides competition!

After peace, we must continue to cultivate culture deeply.
The resources in the world are limited.
Only wisdom can be added.
I want to increase the warmth in my heart,
I want to extend the warmth to life.

Affect

Yesterday, I was in a bad mood.
The tone of speech was very rude.
I feel very good today,
The tone of speech also becomes soft,
Dreams affect reality.

Culture is accumulated,
What clothes do people wear?
What kind of house do people build?
Do people value the environment important?
What do the big trees think about?

Cultural Root

I used to be a blank sheet of paper.
I don't know "Who am I?"
I don't know what I like,
Do not appreciate, do not cherish, do not think,
There is no sound mind.

The big tree was once a blank piece of paper.
Who are we?
Is there a culture now?
What is the new culture called?
Does the big tree have a cultural root?

Green City

How to build a beautiful city?
How to build a green city?
Big trees have a long life,
"What should I leave for tomorrow?"
The big tree often thinks about it.

Thinking about tomorrow,
Predicting a happy future,
Thinking about it and there was a blueprint.
The big tree also saw the sun in the heart.
The big tree really has a culture.

Kingdom of Dreams

How to build a beautiful city?
The big tree tells everyone the blueprint,
This is culture,
We have a common belief,
Life has the proper posture.

The desire has gradually come true,
Mind, language, action,
Clothes, house, city,
From the inside out,
From the kingdom of dreams to reality.

The Eighth job

Only the light can defeat the darkness,
Only the fruit can stop hunger,
Peace is not empty,
Culture exists in every corner.

What is true peace?
Warm big tree,
Warm me,
We have a common belief.

To build a beautiful city,
This is the big tree's and my work.

Commitment

The big tree has a new culture,
Culture is the attitude that life should have.
It is the answer to the meaning of life,
What is the new culture?
The sun in the heart is the answer.

What to do after peace?
Culture needs deep cultivation,
Culture should not be just a blueprint.
Life is full of civilization,
There are a lot of things to do.

Journalist

On an ordinary day,
Eating bread,
Can I trust the baker?
This is culture,
If you don't have faith, there is no bread.

On an ordinary day,
Reading the newspaper,
Does the journalist have the spirit of professionalism?
This is culture,
If there is no conscience, there is no real news.

Software

On an ordinary day,
Turn on the light,
It's really convenient!
This is culture,
Humans are no longer afraid of lightning.

On an ordinary day,
Playing a little game on my computer,
The computer has hardware and software.
This is culture,
Knowledge is not intangible.

Friendship

Because the big tree has fallen leaves,
Sweeping leaves forms a job,
Now the tree is moving in a more civilized direction.
What does more civilization represent?
How should I do it?

I am already a friend with the big tree.
We will continue to be friends in the future.
Ah! This is also a busy matter.
Friendship needs to be maintained,
Civilization needs to be maintained.

Postman

Friendship needs to be contacted.
Make a call, write a letter,
Invisible friendship turned into words,
The letter is sent to a distant place.
The postman transported friendship.

Culture needs to be cultivated,
Mom said, the teacher said,
The same words have been circulating,
The letter is sent to the future,
Culture has a spokesperson.

Specific

Culture is spreading,
When cutting hair,
The hairdresser will care about my current situation.
Culture is passed on,
The nurse's hands are very warm.

Thanks doctor,
This is not the relationship between consumers and industry.
The teacher is not just teaching,
The owner of the grocery store is very kind,
Culture is this specific.

Keep Smiling

The bus driver is serving for me.
I paid the ticket,
I also said "thank you".
This is my duty,
It is my duty to respect others.

Work is to serve others,
I keep smiling,
Even if my life is not complete,
I can still care about the people around me.
Smiling is part of my job.

The Ninth Job

The cashier at the bakery has a sweet smile.
She made the bread delicious.
This is culture!
Culture is delicious!

Eating bread,
I gained strength,
It's my turn to serve others,
I want to bring smiles to others.

Bring a smile to the big tree,
This is my job.

Lady

It's a new day,
"What are you going to do today?"
What does the big tree think of in its heart?
Sunlight enters the soil,
The things that I think are turned into behaviors.

The culture has become concrete,
Culture has become a gentleman and a lady.
Accumulated day by day,
People left their footprints,
The culture has become more specific.

Mobile Phone

The culture has become concrete,
The library has a lot of books.
Wisdom is accumulated,
The library bought a computer,
The computer can keep more books.

The culture has become concrete,
I have a mobile phone,
I can go online,
What books are there in the library?
Check it out online!

Pigeon

I am surrounded by technology,
There are farming techniques in the fruit.
There is textile technology inside the clothes.
I have been vaccinated against diseases,
The taste of modern people is different from that of primitive
people...

There are several courtesy umbrellas at the station.
Courtesy umbrella has not been stolen,
There are pigeons in the park.
Pigeons are not afraid of people,
There are fish in the river.

Inert

How do you cultivate culture?
The big tree once did not want happiness.
Now there is the sun in the heart of the big tree,
But it still need to practice,
Building a good habit is hard work.

I also hope that the environment is clean and tidy.
But it's so troublesome!
The trash can is far away.
A wish is not a fantasy,
I have a reason for self-requirement.

Motive Power

Culture is like this,
Others are making progress,
Can I not improve?
Although I am inert,
I still have to change.

Culture is like this,
Culture is not a rigid rule.
There is a real reason,
There is a driving force for progress,
I know why I have to move forward.

Inventor

People want better life,
People are very lazy,
Ah! Troublesome!
Is there any good way?
Is there any way to be lazy?

The tree is very lazy,
The customer is very lazy,
I am very lazy,
The ideal and inertia sparked,
The inert me became an inventor.

Delivery Staff

Ah! So lazy,
Even lazy to go out to eat,
I really hope someone can send dinner to my house.
Thanks to the delivery staff,
My wish has come true.

People want the world to be better,
People are willing to do their part,
But people can't think about big morals,
Ok, give it to me,
Let me make culture simple.

The Tenth Job

The library has a lot of books.
Too many,
A bit messy since there are too many,
The administrator arranged the books neatly.

I found the book I wanted,
I have a sound mind,
I have my profession,
I want to make my profession simple.

Make something difficult easy,
This is my job.

Concentration

Time continues,
The big tree accumulates wisdom,
The problem is getting easier,
Wisdom condensed into common sense,
Wisdom condensed into a habit.

Wisdom is accumulated,
Wisdom is again concentrated,
Does the big tree have a sound mind?
Is there enough wisdom?
Is it enough to form a self?

Belief

The meaning of life is blank,
The blank is still too abstract,
What is happiness?
Is happiness also abstract?
When can I have a definite answer?

What is happiness?
Still looking for it?
What is the answer?
Still looking for it?
When do you want to hold happiness?

Conclusion

Learn to cherish things every day,
Repeating possession and loss,
Is there anything that will not change?
Repeating possession and loss,
The heart has been learning to cherish.

The big tree has a sound mind,
The big tree knows the heart,
Already clearly seen,
Already confirmed,
The answer is there.

Consensus

Sweet? Bitter?
Tasty?
What is the taste of "Today"?
What is the taste of "I"?
The big tree bears fruit.

Who am I?
The tree knows the heart,
Everyone knows the heart,
Culture has a name,
That name is "the consensus of heart."

Our Name

Invisible heart,
Tangible heart,
Big tree's heart,
Beautiful earth,
We are one.

The name stringed everything up,
Dream, body and matter,
You, me and the big tree,
Big life has a name,
Big life is a beautiful heart.

ID

Why can money buy things?
Because we are one,
The name of the country is printed on the banknote.
Exchange of goods and services,
We have to cooperate.

The household clerk confirms the name,
Bank personnel confirms the name,
I took out my ID,
The name of the country and my name are on the ID.
The owner of wealth is the name.

Discovery

"The world is beautiful!"
Artists said so,
"The world is wonderful!"
Scientists said so,
I confirmed the name again.

You are a beautiful heart,
The big tree is a beautiful heart,
I believe I am, too.
I have repeatedly discovered the beauty of the world,
This is a busy thing.

The Eleventh Job

The flower appeared,
The flower has disappeared,
Flowers and people playing hide and seek,
This is an eternal game.

Found it,
I confirmed,
Cherish precious names,
This is the responsibility forever.

There is a name on the banknote.
Polishing the name, this is my job.

Resolution

Time passes,
The characters in the game are not getting older.
The skin of the video game character has become smoother.
Civilization is getting more and more progressive,
The resolution of the game has improved.

Time passes,
The big tree depicts the heart,
The big tree depicts today,
The big tree depicts itself,
The heart is more and more specific.

Destiny

What are you going to do today?
What have you done today?
The invisible heart became yesterday,
The invisible heart becomes a footprint,
The invisible heart becomes a trajectory.

Yesterday becomes the flesh and blood of the big tree.
Destiny stands specifically in front of me,
The fate of light is still a destiny.
I am constrained,
Happiness tells me to cherish happiness.

Home

One day, a little dream came true.
I have a home,
She is in front of me,
The answer is completely fixed,
She is my answer.

One day, another dream came true.
I became friends with the big tree,
It's a warm country in front of my eyes,
The answer is completely fixed,
The sun illuminates the world.

Thank You

The answer is always in front of me.
Eating with her,
Big tree that encouraged me,
The eternal sun,
Truth is always in front of my eyes.

The answer is always in front of my eyes,
It's only one step away from happiness.
Just saying it is left,
Thanks for everything,
I want to say my thanks.

Important Position

Life is meaningful,
The earth is precious,
But people have never realized it,
Happiness is not happiness,
Cherishing happiness is happiness.

There is a sanctuary in my heart.
The big tree has a sanctuary.
What does the big tree cherish most?
Does the big tree have an answer?
Does it put the answer in the most important position?

Attempt

The gourmet has eaten in a delicious restaurant.
The gourmet has eaten in the unpalatable restaurant.
The gourmet recommends good restaurants.
Edison had thousands of failures,
Edison invented the light bulb.

The work is like this,
Try, fail, try, and fail again,
I taste the failure,
Customers enjoy success,
Finding the answer is my job.

Network

The big tree has complex networks,
The maintenance staff sent electricity to my home.
Not just electric lights,
A lot of things need electricity,
I was tightly wrapped in the power grid.

The big tree has complex networks,
What is the function of electricity?
What is the function of a warm heart?
Trust, friendship, the rule of law, warmth...
The warm heart is covering the big tree.

The Twelfth Job

Can I trust you in front of me?
Are you a civilized person?
You won't suddenly attack me?
Is this a civilized country?

Warm heart is very important,
I am a maintenance person.
I have to say it out loud,
I have to say the name of the heart.

Implement "The Consensus of Heart",
This is my job.

Born

Woke up,
Another new day,
What is waiting for me is not a war,
What is waiting for me is work,
There is something to do today.

Sun, air, water,
Heart, earth, today,
I was born on this earth,
I'm hungry,
I want to work for the big tree.

Tacit Understanding

Another new day,
I'm hungry,
The big tree is also hungry.
I have to eat well today,
I have to be busy for dinner today.

Going to work,
Going to the breakfast shop,
Still the same,
I don't need to tell the boss what I want to eat.
I have a tacit understanding with the big tree.

Old driver

Take the bus,
Still the same route,
Still the same driver,
His driving skills are very good,
He is a skilled old driver.

This is a clean street.
I didn't see the fallen leaves,
Did you see it?
Why are there no leaves?
I saw the hard work of the cleaning staff.

Meeting

Arriving at the company,
There is another meeting,
A meeting that will never finish,
The topics for the meeting are the same:
What products are we going to launch next season.

The boss is always unreasonable,
The boss doesn't want to listen to long instructions.
Customers are always lazy,
Customers only accept convenient products,
I want to make difficult things easy.

Doze off

Boss!
You can't doze off during a meeting.
Even if you are the boss,
If you don't listen carefully,
How can I help you solve the problem?

I serve the customer,
I work hard to show my smile,
Guest!
I hope to come up with the best service,
But my smile has a bottom line.

Service

Today I serve others,
Others will serve me tomorrow.
I really hope that we can respect each other,
There are this many resources on earth.
Only culture is real wealth.

"I hope that people can respect each other"
I want to fulfill this wish,
I pay taxes,
The state has legislators and police,
I hope that the big tree has culture.

Name of Culture

Happiness has outlines,
Happiness is alive,
Respect cannot be written into law in details,
Respect is warm,
I want to polish the light in my heart.

There is support behind the law,
Culture has its roots,
Get to know more about the heart!
"The Consensus of Heart",
This is the name of culture.

Free Time

Technology brings convenience,
Thank you for the sun in the heart,
Technology did not bring destruction.
The work has yielded results,
People have more free time.

I have some time,
The big tree has some time,
How to fill up the time?
I got a little time through hard work,
I didn't expect time to push me to work.

Holiday

It's too busy to have a holiday.
To accompany parents,
To accompany the wife,
To accompany the child,
To contact old friends.

What to do during the holiday?
Prepare a dinner party,
Prepare a small trip,
This is also work!
Maintaining friendship is a busy matter.

Exercise

It's vacation time,
Having the free time,
Ah, I haven't exercised for a long time.
This is also my job,
Maintaining health is my job.

I want to exercise,
Not because of hunger,
Not because of fallen leaves,
Exercise is because of time,
Time pushes me to exercise.

Return Home

After the exercise, returning home,
Getting back to my own room,
Finally, there is really time for myself,
But another job has appeared,
I have to arrange my time.

I want to entertain myself,
I want to enrich myself,
This is work,
This is the responsibility that cannot be pushed aside.
Time is waiting for me to fill in.

Lie Down

Flipped over some books,
Read some comics,
Took a shower,
Lying in bed now,
Finally calming down.

Time!
The sun!
This is another job,
I have to take care of my heart,
Time itself is work.

Need

Time itself is work,
The big tree has a long life,
The big tree needs people's services,
I have to take care of the big tree.
I have to take care of the heart of the tree.

The work is like this,
I don't want to work!
But the leaves always need someone to sweep,
Actually it is not what I want to do,
But the big tree has demands.

Hunger

"What to do today?"
I get hungry,
The big tree needs to be taken care of,
Today needs to be filled,
There are things to be done for today.

Heart, the Earth, today,
I am hungry,
The big tree is hungry,
Today is hungry,
I must present a dish.

Grumpy

The invisible heart is hungry,
The tangible heart is hungry,
My family is hungry,
They need food,
They need accompany.

I want to eat a meal with laughter,
What to do?
What should I do?
Big tree!
You can't act grumpy.

Work for the Big Tree

Money comes from the big tree,
If the big tree is not normal,
I can't eat well, too.
Ok! This is my job,
I want to remind the big tree.

Big tree!
You have an obligation,
You have to obey the civilization,
You have to know the truth,
You have to know the heart.

Movie

Can I choose not to work?
I really want to lie on the sofa,
Drink cola, eat potato chips, watch movies,
What to eat for dinner?
Call for a pizza!

It's a good movie!
Can't help but stand up,
Thank you, director, thank you actors,
Hard work,
You are so great.

Allocation

Can I choose not to work?
Scientists invented the machine,
Great, life is getting easier,
What to do?
Many people are unemployed.

The bread has become more and more.
There are fewer people who are full,
What to do?
The state will not take the initiative to care for the people,
Remind the big tree is our job.

Irreplaceable

Can I choose not to work?
Do I have to let the robot feed me?
The robot is the robot,
The robot won't eat, poop,
The robot does not understand the joy of eating.

Exercise is to maintain healthy,
Accompanies are to maintain family bonds,
My hands are warm,
Some jobs can only be done by me,
I am irreplaceable.

My Work

Can I choose not to work?
The sun asked me:
"What are you doing today?"
This is work,
I have to arrange today.

Arranging today,
Knowing myself,
I understand the hints of the sun,
I have to take care of my heart,
This is my job.

My Heart

I used to be a blank sheet of paper.
Can't talk, can't write,
Don't know the word "I",
I don't know what I like,
I am not responsible for myself.

Now that I am growing up,
Filled in the answer in the blank.
I found something important,
I found a heart around a circle,
My heart, her heart, the heart of the big tree.

The Bright Future

The tree has had a dark age,
Life doesn't make sense,
People don't know what to do today,
No future direction,
So they did stupid things.

What are you going to do today?
What to do in the future?
The sun is shining,
Ah! This is the answer!
The sun shines on the heart of the big tree.

Meaning of Life

Does life have meaning?
What should people do when they are born into this world?
I will be hungry,
The big tree needs care,
Today is waiting to be filled in.

This is homework,
This is work,
Life has meaning,
There are things that must be done today,
The meaning of life is by eating
Know the blank heart.

Afterword

◎Culture has a process and roots.
　　But politicians only want convenient slogans.

◎The root of culture is the sun,
　　And the sun in my heart.

◎Thanks to the sun,
　　Thanks to the beautiful heart.

◎The person who knows the answer,
　　Has the obligation to tell people who
　　do not know yet.

為大樹工作（國際英文版）

Work For The Tree

作　者/決長（Jue Chang）

作者專頁/www.amazon.com/author/jue.chang

出版者/美商 EHGBooks 微出版公司

發行者/美商漢世紀數位文化公司

臺灣學人出版網：http://www.TaiwanFellowship.org

印　　刷/漢世紀古騰堡®數位出版 POD 雲端科技

出版日期/2019 年 11 月

總經銷/Amazon.com

臺灣銷售網/三民網路書店：http://www.sanmin.com.tw

三民書局復北店

地址/104 臺北市復興北路 386 號

電話/02-2500-6600

三民書局重南店

地址/100 臺北市重慶南路一段 61 號

電話/02-2361-7511

全省金石網路書店：http://www.kingstone.com.tw

定　　價/新臺幣 450 元（美金 15 元 / 人民幣 100 元）

www.ingramcontent.com/pod-product-compliance
Lightning Source LLC
Chambersburg PA
CBHW031304060726

47590CB00003B/1048